THE JAMES WEBB SPACE TELESCOPE

"A New Era of Astronomy" How NASA's latest and most advanced space observatory. It's revolutionizing our grasp of the universe.

Brian S.A. Corey

02

Copyright Note

All rights reserved. No part of this publication may be reproduced, distributed, or transmitted in any form or by any means, including photocopying, recording, or other electronic or mechanical methods, without the prior written permission of the publisher, except in the case of brief quotations embodied in critical reviews and certain other noncommercial uses permitted by copyright law.

Copyright © Brian S.A. Corey, 2024.

Disclaimer

This book- is titled "The James Webb Telescope." It aims- to provide accurate- and authoritative information- on the subject. The author, Brian S.A. Corey, has made every effort- to ensure the information in this book is correct. The author- is not responsible for any- loss, damage, or disruption caused by errors- or omissions in this book. These errors or omissions could result from- negligence, accident, or any other cause. Please note that- the James Webb Space Telescope and its research- are constantly changing. New discoveries- may replace or improve the information- in this book. The author's views and opinions- in this book may not represent the official- policy or position of any U.S. government- agency or organization.

NEPTUNE'S RINGS	CARTWHEEL GALLERY
PILLARS OF CREATION	CARINA NEBULA
NGC 346 STAR CLUSTER	COSMETIC DANCE

About The Author

Brian S.A. Corey is an American author and astronomer. He has always been interested in understanding and exploring the universe. The title of his self-published book is called "James Webb Telescope." It goes into detail about the James Webb Space Telescope and explains why it was created. It also shows how it will greatly improve our understanding of the universe.

Brian takes complicated scientific topics and turns them into captivating stories. This has earned him recognition in both the literary and scientific communities. His work excites readers, sparks their interest, and encourages an interest in astronomy.

Brian is a writer and researcher who is involved in public outreach projects. He contributes to astrophysics and space research. Join Brian as he explores space and time with the James Webb Telescope. Discover what the future holds for astronomy research.

TABLE OF CONTENTS

INTRODUCTION — 09

Chapter 1 — 13
Everything You Need To Know About James Webb Space Telescope

Chapter 2 — 33
Design and Development

Chapter 3 — 41
James Webb Telescope Snaps Stunning Shot of Uranus

Chapter 4 — 57
Launch and Deployment

Chapter 5   **73**

First Images of Mars Captured by
James Webb Space Telescope

CONCLUSION   **95**

INTRODUCTION

Have you ever wondered- how the cosmos came- to be? Did you ever wonder how it changed or how it will end? Have you ever imagined what the earliest stars and galaxies could have looked- like? Did you consider how they formed the universe- and what mysteries they hold? Have you ever searched for indications of life on other planets? Did you look- through the billions of planets orbiting other stars?

If you have, many others- have too. For centuries, people have been interested- in the mysteries of the universe. They have used different tools and methods to explore them. The JWST is one of the most powerful- tools ever made for this. NASA, ESA, and CSA worked together- to create it. It was sent into space in 2022. The JWST is a space- observatory.

It studies the universe in the infrared- range of light. This helps it see new details and features- that other telescopes couldn't see before.

The JWST is a telescope- and a time machine. It can look back to the beginning- of the universe, when light came after the Big Bang. It can see stars being born- and dying, galaxies forming- and changing, and planets- being made- and destroyed. It can also- look for signs of life on exoplanets, which are planets- that go around stars that are not- our sun.

The JWST is more- than just a time machine. It is also a discovery- machine. It can answer important questions in astronomy- and cosmology:

- How- did the- first stars form?
- How did the first- galaxies form?
- How did- they affect the space between- them?
- How did the- elements that make up life- originate?
- How did they spread- throughout the universe?

How many planets- that can support- life are there around other stars? Are we the only- ones in the universe? The JWST is not just a machine- for making discoveries, but also an amazing- engineering and science- achievement.

10

It has a 6.5-meter mirror, the largest ever flown- in space, that can collect more light- than any other telescope. It has four- scientific instruments. Each has its own unique- abilities and functions. They can take pictures and measure the infrared range.

It is positioned- at the second Lagrange point (L2), about 1.5 million kilometers- from Earth. This location- protects it from the bright interference- caused by the sun, Earth, and moon. It has a view of the entire- sky, allowing it to observe- any part of the sky at any time.

The JWST is a marvel- of engineering- and science. It is a legacy of human- curiosity and creativity. It took- decades to plan, design, build, test, and- launch. People from different- countries, disciplines, and backgrounds worked- on it.

The Hubble Space Telescope- (HST) came before it. The HST changed our understanding- of the universe with its amazing images- and discoveries. The JWST is a preview of what future- missions will do. They will expand our horizons- and knowledge.

The JWST is more than- just a result of human curiosity- and creativity. It is a gift- to all of humanity and future generations.

It inspires and amazes us. It educates and connects us. It encourages innovation and collaboration. It gives us hope and something to strive for.

The James Webb Space Telescope / Brian S.A. Corey

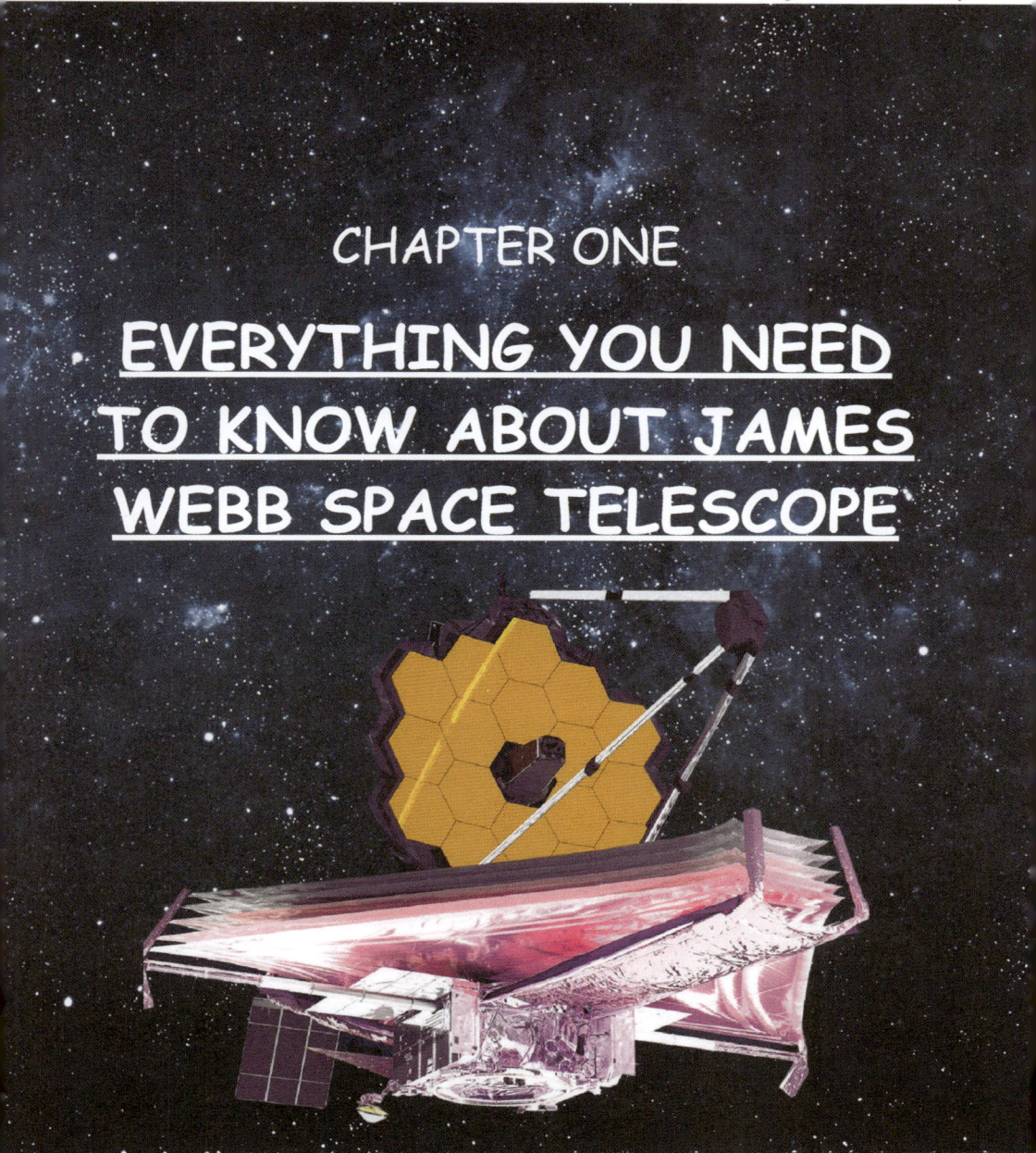

CHAPTER ONE
EVERYTHING YOU NEED TO KNOW ABOUT JAMES WEBB SPACE TELESCOPE

Functioning as an infrared observatory, the James Webb Space Telescope improves and enlarges the accomplishments of the Hubble Space Telescope. It provides an expanded wavelength range and notably heightened sensitivity.

In 2002, James E. Webb was honored by NASA with the naming of the James Webb Space Telescope. Webb oversaw NASA from 1960 to 1968 and was a key figure in the Apollo, Mercury, and Gemini missions. He focused on human spaceflight and space science. Webb pushed for a big space telescope, which later became the Hubble Space Telescope. Naming its successor after Webb was a way to recognize his contributions to space exploration.

The JWST is an observatory. It studies the history of the universe. It is located 1 million miles from Earth. The telescope observes the earliest stages of the universe, star formation, and more. It was launched on December 25, 2021, using an Ariane 5 rocket from Europe's Spaceport. On January 8, 2022, the telescope's primary mirror was unfurled. Its journey ended on January 24, 2022. It reached its final destination: orbiting the sun at the L2 Lagrange point.

Webb can see a lot of the cosmos from this location. The location also keeps the scientific instruments at the right temperature. On July 11, 2022, President Joe Biden showed a picture from the telescope. The picture is in color and shows a galaxy cluster called SMACS 0723. The picture is called Webb's First Deep Field. It shows the faraway universe in infrared. You can even see the faintest objects in the cluster.

The James Webb Space Telescope / Brian S.A. Corey

The photos- in Webb's First Deep Field were captured using Webb's Near-Infrared Camera (NIRCam). These photos were captured- during the course of 12.5 hours. The depth of these photographs is larger than the depth of Hubble Space Telescope infrared images. The Hubble photos- were captured over a period of many- weeks. The picture- depicts the 4.6 billion-year-old galaxy cluster- SMACS 0723.

The next- day, July 12, 2022, further

On July 12, 2022, the following- day, additional images- were unveiled. They showcased the capabilities of all four advanced scientific instruments aboard Webb.

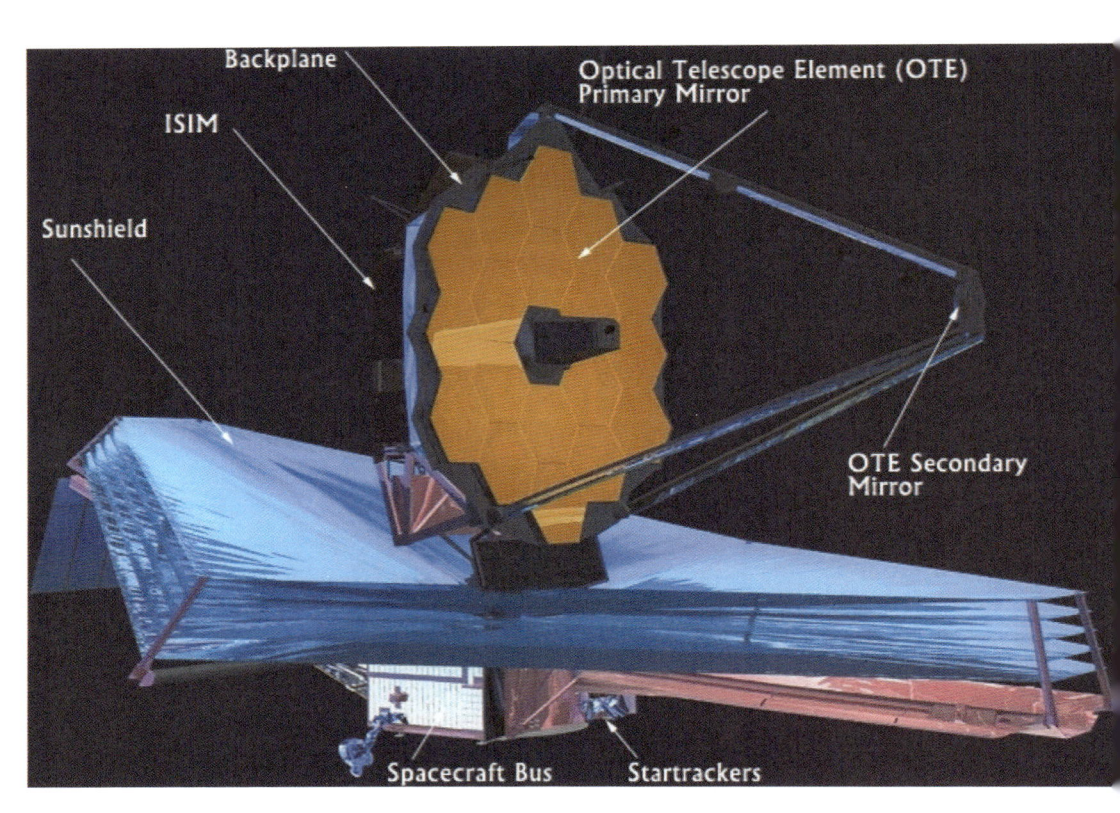

Webb's instruments- are essential to the James Webb Space Telescope- Observatory's flight system. They are housed in the Integrated Science- Instrument Module (ISIM). The ISIM is one of the three- crucial elements. The other- two are the Optical Telescope Element (OTE) and the Spacecraft- Element (Spacecraft Bus and Sunshield).

The James Webb Space Telescope / Brian S.A. Corey

Inside the enclosure is the Integrated Instrument Module (ISIM). It contains the science instruments NIRCam, NIRSpec, MIRI, FGS)

Momentum Flap

Spacecraft Bus

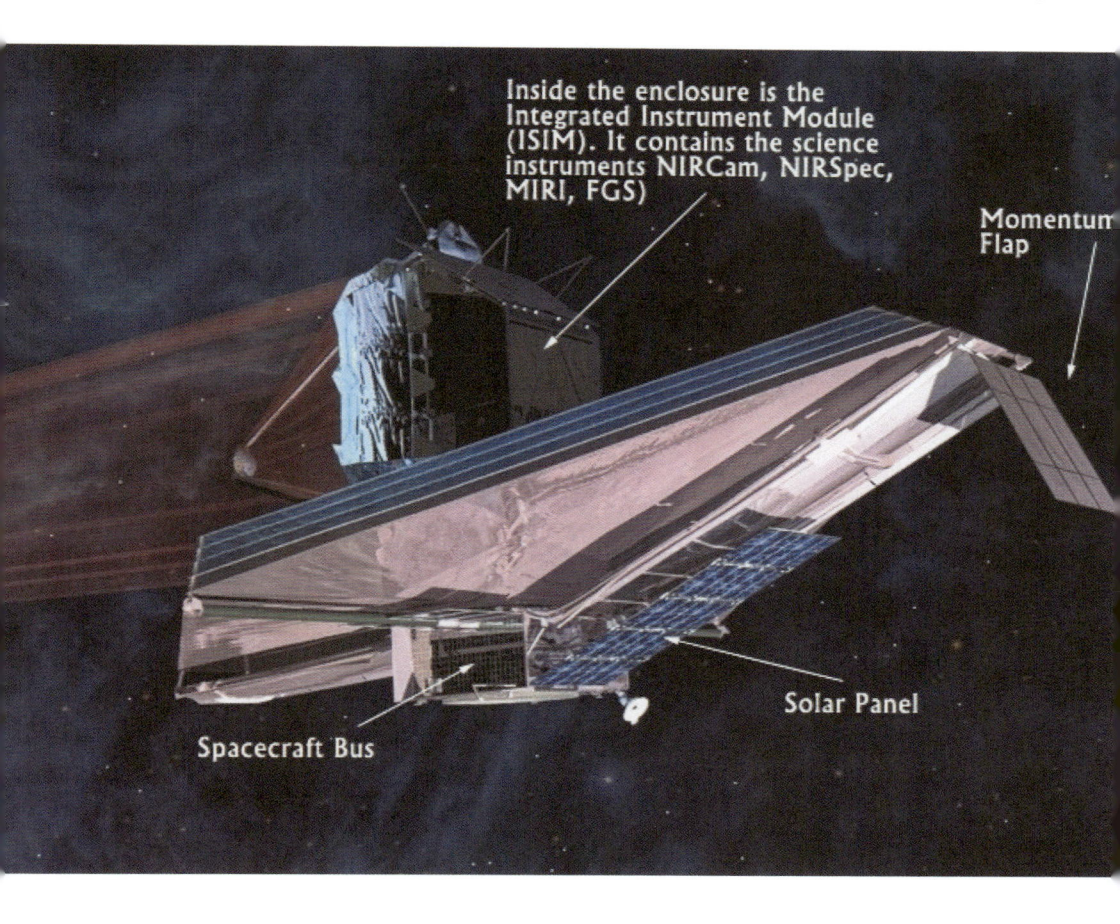

Solar Panel

The ISIM is the main- part of the James Webb Space-Telescope. Engineers call it the primary payload. It has four primary instruments. These instruments capture light from faraway stars, galaxies, and exoplanets.

The ISIM includes these instruments:

·The University- of Arizona supplies the Near-Infrared Camera, which is abbreviated- as NIRCam.

- ·The Near-Infrared- Spectrograph, called NIRSpec, was provided- by the European Space- Agency (ESA). It includes- components- from the NASA- Goddard Space- Flight Center (NASA/GSFC).

- ·The MIRI is a Mid-Infrared- Instrument. It is provided by the European- Consortium in collaboration with the ESA and JPL.

- ·The Canadian Space Agency- delivered the Fine Guidance Sensor/ Near InfraRed Imager- and Slitless Spectrograph, also known- as FGS/NIRISS.

ISIM Components within the Observatory

Region 1: Instruments are mounted to ISIM structure and enclosed by observatory enclosure and radiators.

Region 2: ISIM Electronics Compartment (IEC), provides mounting surfaces and ambient thermally controlled environment for instrument electronics in close proximity to instruments

Region 3: Spacecraft houses ISIM Command and Data Handling (ICDH) and cryo-cooler compressor and cryo-cooler electronics

IN-DEPTH: The ISIM has many technical insights.
It is a big challenge to put four important instruments and many subsystems together in one payload. Engineers have divided the ISIM into three separate regions to make integration simpler.

The "region 1" segment cools the detectors to 39 K. This cools them enough to detect distant cosmic infrared light, which is essentially heat. The ISIM/OTE Thermal Management Subsystem cools passively. Other devices cool the detectors even more.

The "Region 2" section is the ISIM Electronics Compartment. It provides a controlled thermal environment. It also has mounting surfaces for the instrument control electronics.

The "region 3" of the Spacecraft Bus includes the ISIM Command and Data Handling subsystem. It has the ISIM flight software and the MIRI cryocooler compressor and control electronics.

SMACS 0723: Webb has captured a highly detailed infrared image of the distant universe. It took- only 12.5 hours to create- this image.

To give you an idea of its size, if you were on Earth, it would be like looking at a grain of sand held at arm's length. This image reveals some of the farthest galaxies ever seen by using a lensing galaxy cluster. However, there is much more for Webb to discover about deep fields and the origins of galaxies.

WASP-96b (spectrum): Webb observed an exoplanet outside our solar system. Webb found a clear water signature on the exoplanet. Webb also found signs of haze and clouds that previous investigations had missed. This is the first time water has been discovered in the atmosphere of an exoplanet. Webb will now study hundreds of other systems to learn about the atmospheres of other planets.

Southern Ring Nebula: This planetary nebula is about 2,000 light-years away. It is an expanding gas cloud that surrounds a dying star. Webb can see a second fading star that was hidden before. Webb can track the dust and gas that aging stars release as they become planetary nebulae. Webb can see if these materials eventually become new stars or planets.

Stephan's Quintet: Webb sees inside a compact galaxy cluster in Pegasus. It sees through the dust around a galaxy's center. This reveals information about the gas's speed and makeup near the supermassive black hole. Researchers now have a detailed view of how interacting galaxies cause star formation in each other. They also see how the gas in the galaxies is being disrupted.

Carina Nebula: Webb studied the "Cosmic Cliffs" in the Carina Nebula. This revealed the early stages of star formation that were hidden before.

Webb examined this region of star formation in the Carina constellation and other similar regions. Webb found new stars that are just starting to form and studied the gas and dust that created them.

Here are 12 facts about- the James Webb Space Telescope:

- The space telescope is the biggest- and most powerful one ever built. Scientists can- use it to look into a universe that is about- 200 million years after the Big Bang.

- It works in infrared light, so it can go through dust clouds and find stars and planets that are hidden. It can even find things that are so far away that their light has changed from visible to infrared. This is because the universe is expanding.

- The telescope has a sun shield. The shield- is as big as a tennis court. The shield protects- the telescope's mirrors- and instruments from the Sun's heat and bright light. The sunny part- of the shield is more than 600 degrees Fahrenheit hotter than the shaded part.

- This arrangement consists of 18 smaller mirrors that are intricately aligned. The mirrors are- arranged like a puzzle. Together, they form- a very large mirror with a diameter- of 6.5 meters (21.3 feet). These mirrors- are made from- beryllium coated with gold. This allows them to reflect light very well.

- The telescope is large. It must- be folded, like origami, to fit- in the rocket- that will launch it into space. Once in space, the telescope- will unfold, and the sun shield- will be the first part to open.

- The Sun is circled by a satellite that is about 1 million miles away. The satellite is positioned at the L2 point. This position allows the satellite to stay in sync with both the Earth and the Sun.

- The Webb satellite has four instruments. These instruments are a camera, a spectrograph, an instrument, and a sensor/imager. Webb also has a spectrograph. These tools help Webb study galaxies and exoplanets in space.

- It has a focal length- of 131.4 meters (431 feet). It can focus- on distant or minuscule objects.

- The Webb telescope can study objects with different temperatures and compositions. It can do this because it can detect- both visible and mid-infrared light. It can detect- light with wavelengths between 0.6 and 28.5 microns.

- The mission- lasts 5 to 10 years, depending on how much fuel is used to keep the spacecraft- in the right orbit and point it accurately.

- The total payload mass is approximately 6200 kilograms (13670 pounds). This includes the observatory, on-orbit consumables, and the launch vehicle adaptor.

Here are 10 differences- between the Hubble and James Webb telescopes- from web search results:

- Hubble- observes primarily- in the optical and ultraviolet- wavelengths. Webb observes primarily- in the infrared wavelengths.

- Hubble has a 2.4-meter diameter- mirror. Webb has a 6.5-meter diameter mirror- composed of 18 segments.

- Hubble orbits Earth about- 570 kilometers above. Webb orbits- the Sun, positioned- at the L2 point 1.5 million kilometers from Earth.

- Hubble has four instruments. The instruments are:

- A wide field camera
- A cosmic origins spectrograph
- An advanced camera for surveys
- A space telescope imaging spectrograph.

- Webb also has four instruments. The instruments are a near-infrared camera, a near-infrared spectrograph, a mid-infrared instrument, and a fine guidance sensor. The fine guidance sensor also works as a near-infrared imager and slitless spectrograph.

- Hubble's focal-length is 57.6 meters. Webb, on the other-hand, has a focal-length of 131.4-meters.

- Hubble- covers wavelengths- from 0.1 to 2.5 microns. Webb covers wavelengths- from 0.6 to 28.5 microns.

- Hubble's mission has lasted- over 30 years- and is still- ongoing. Webb's mission- lasts 5 to 10 years.

- The space- shuttle launched- Hubble in 1990. Astronauts serviced Hubble- five times. An Ariane-5 rocket will launch Webb in 2021. Astronauts will not service Webb.

- Hubble can see galaxies that formed 500 million years after the Big Bang. Webb can see galaxies that formed 200 million years after the Big Bang.

- Hubble weighs 11,000 kilograms. Webb weighs 6,200 kilograms.

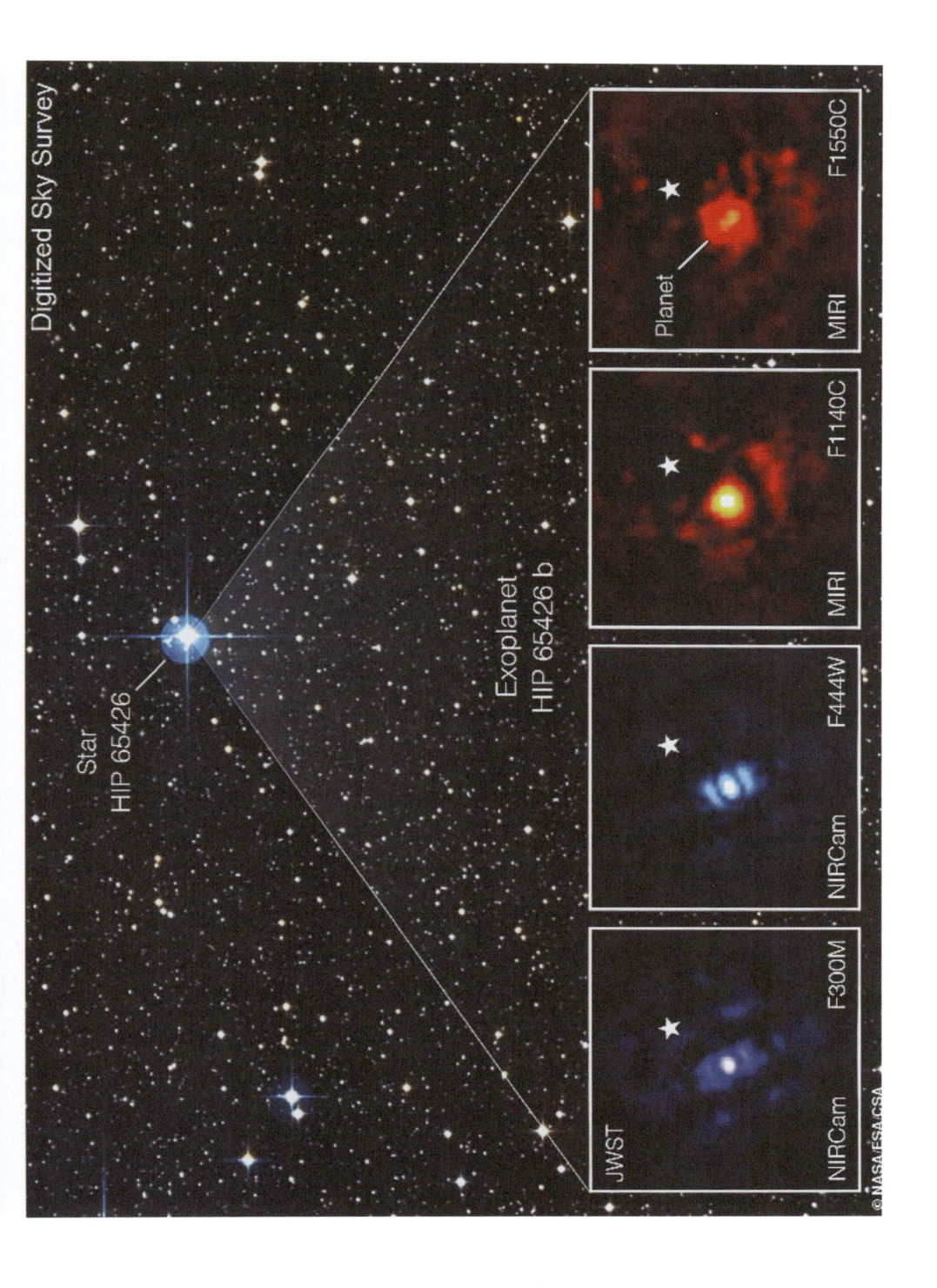

The James Webb Space Telescope / Brian S.A. Corey

NGC 346 is a very active area where stars are forming in nearby galaxies. The NASA/ESA/CSA James Webb Space Telescope has made new discoveries that have helped us understand this area better.

The James Webb Space Telescope / Brian S.A. Corey

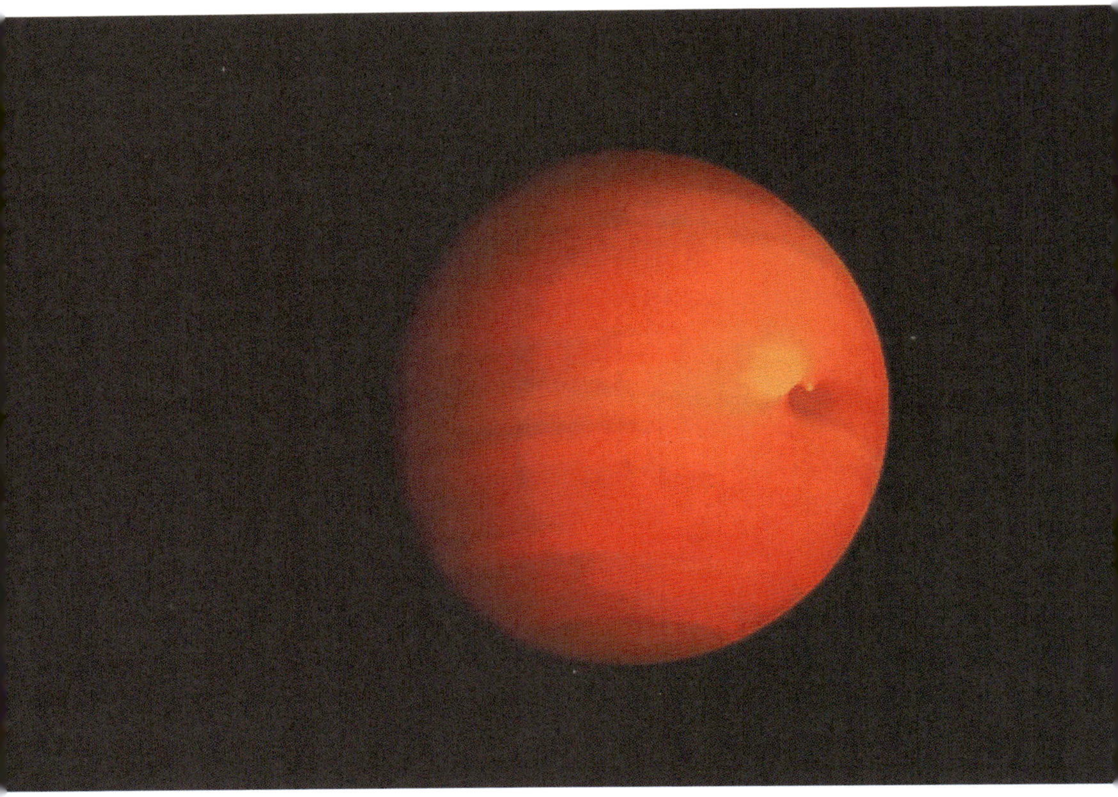

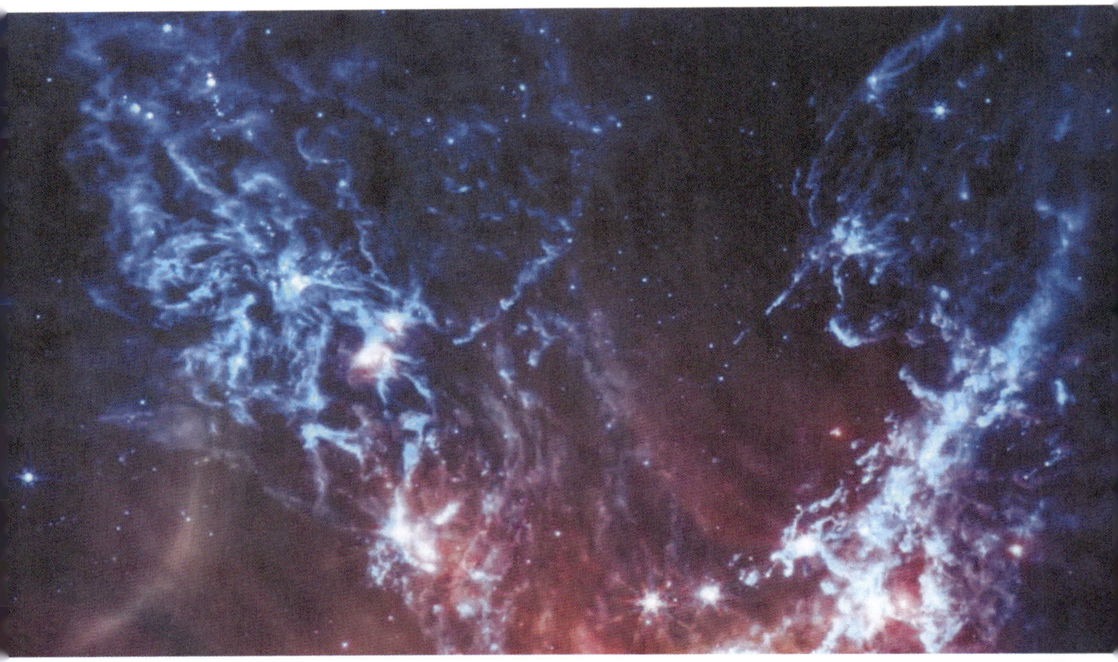

The James Webb Space Telescope / Brian S.A. Corey

main disk plane

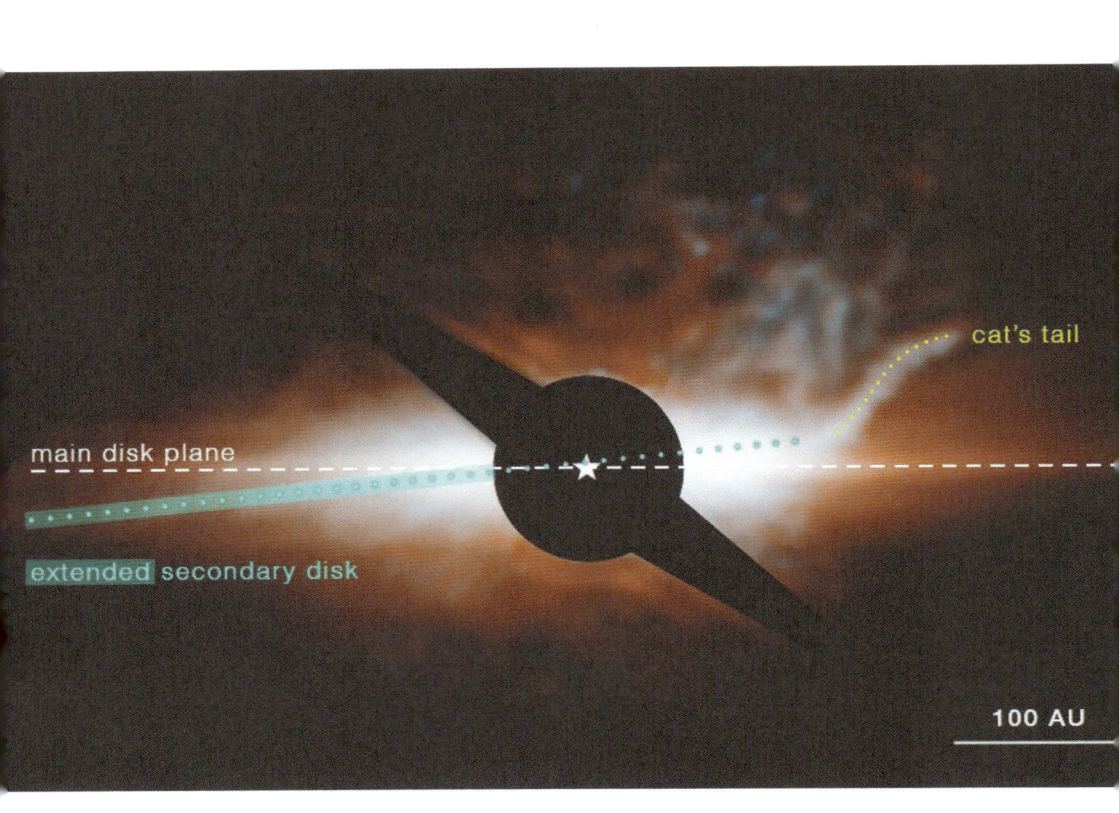

extended secondary disk

cat's tail

100 AU

The James Webb Space Telescope / Brian S.A. Corey

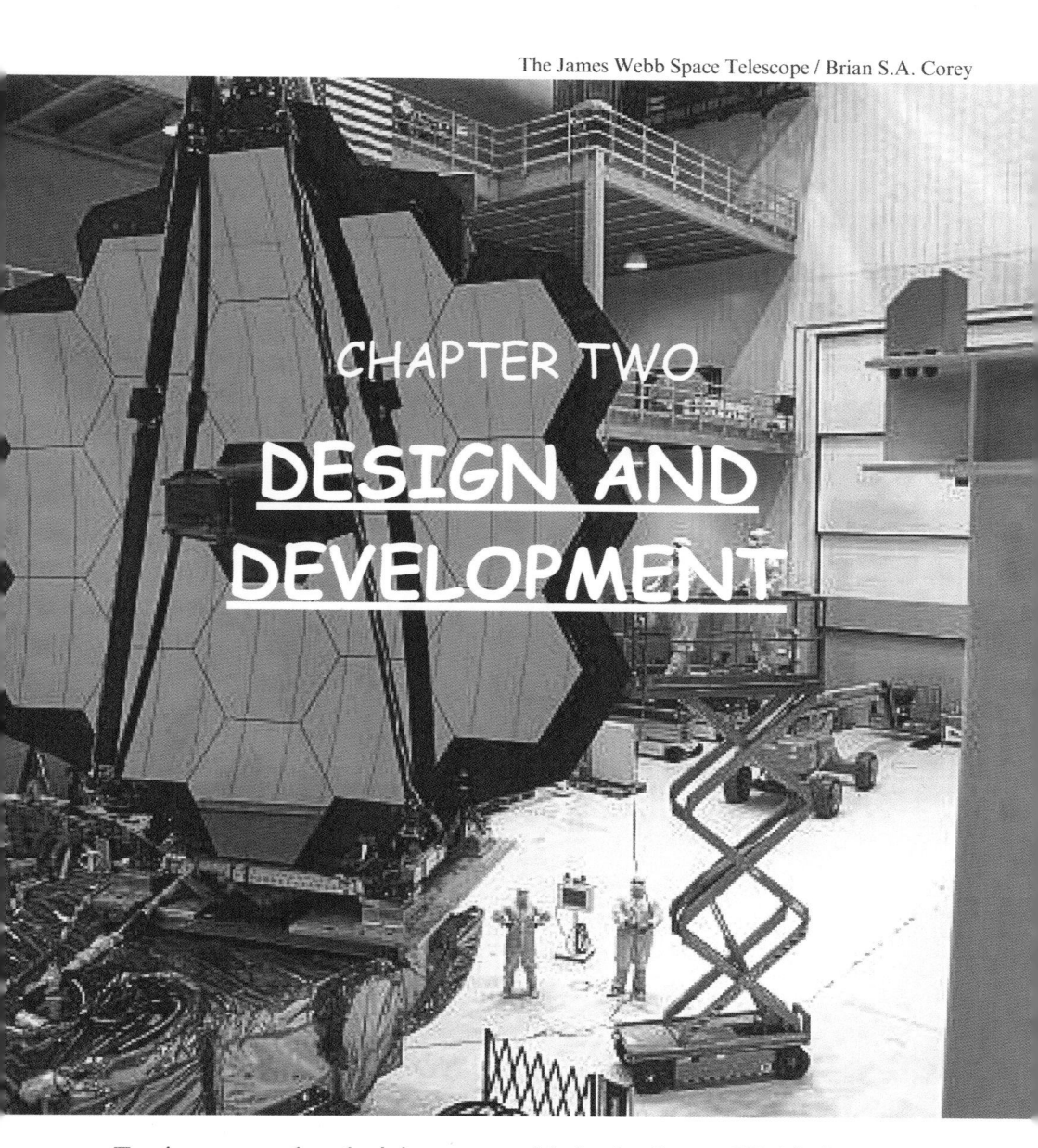

CHAPTER TWO
DESIGN AND DEVELOPMENT

Engineers- and technicians assembled- the James Webb Space Telescope at NASA's Goddard- Space Flight Center. It's located- in Greenbelt, Maryland. They did- this on November 2, 2016. It will- replace the Hubble Space- Telescope and the Spitzer Space Telescope- as a large space-based observatory. It is designed- for infrared wavelengths. The new telescope is scheduled- for release in October 2018.

What are the main components and features- of the telescope

The James- Webb Space Telescope- (JWST) is an impressive- achievement- in engineering- and science. It has many- parts and features that make it a strong and flexible observatory. Here are some- of the main ones:

The primary mirror: The JWST has- a 6.5-meter primary mirror. It is made of 18 hexagonal- segments of gold-coated beryllium. The mirror can collect- more light and achieve higher resolution than other space telescopes. The mirror segments can be adjusted- individually to form a perfect shape and focus.

The sun shield: The sun shield- protects the telescope- from the Sun, Earth, and moon. The telescope stays- very cold, about -223°C or -370°F. This is important- because it helps the telescope- see faraway and dim objects. These objects give off infrared radiation.

The instruments: The JWST has four scientific instruments. These instruments are located behind the primary mirror.

The JWST orbits the Sun at the second Lagrange point (L2). L2 is about 1.5 million kilometers (930,000 miles) from Earth. At L2, the JWST can see the whole sky without interference from the Earth's atmosphere. However, this makes it harder to communicate and fix. If something goes wrong, it's harder to repair.

How was- it designed, built, and- tested?

The JWST is an amazing- achievement in engineering and science. It took- a team of thousands of experts- from NASA, ESA, CSA, universities,- and industries over two decades- to design, build, and test it.

The design- of the JWST was influenced- by the mission's scientific- goals and targets. These goals include studying- the beginning and development of the universe. It will study- everything from the first light after the Big Bang to the formation- of stars, planets, and- galaxies. The design- also took into account- the technical- challenges and limitations- of launching and operating- a space telescope- in the infrared part of the electromagnetic- spectrum.

The JWST- has four main components:

- The optical telescope element- (OTE) has a primary mirror- and a secondary mirror.
- The ISIM has four scientific instruments.
- The spacecraft element (SCE) includes the spacecraft bus and the sun shield.
- ·The launch vehicle element (LVE) comprises of the Ariane 5 rocket, which launches the JWST.

The JWST was built by putting together and combining these parts in different places around the world. The OTE was built- by Northrop Grumman in California, USA. NASA's Goddard- Space Flight- Center in Maryland, USA, built- the ISIM. The SCE- was built- by Northrop Grumman- in California, USA. The LVE was built in France by Arianespace.

Scientists extensively tested the JWST. They wanted to make sure it could survive the tough conditions of launch and work in space. They used different types of simulations.

These- included thermal, acoustic, vibration, shock,- electromagnetic, and optical- tests. The testing also included different levels of integration. These included component, subsystem, system, and observatory-level tests.

They conducted the testing at different facilities around the world. For example, NASA used the Johnson Space Center in Texas, USA. They also used the Plum Brook Station in Ohio, USA. In addition, they used Northrop Grumman's Space Park in California, USA. Lastly, they used ESA's European Space Research and Technology Centre in the Netherlands.

The JWST completed its last testing in August 2021. It was on its way to its launch location in Kourou, French Guiana. The JWST was delivered to South America by water via the Panama Canal. The scheduled launch date was October 31, 2021.

What challenges and risks were involved in the project?

The JWST stands as an immensely ambitious endeavor, seeking to transform our comprehension of the cosmos by observing the most remote and dim entities within the infrared range. However, such a groundbreaking mission also comes with many challenges and risks that could jeopardize its success. Here are some of the main ones:

Technical- complexity: The JWST is made up of many small and delicate- parts that have to work perfectly together in space. For example, the telescope- has a 6.5-meter mirror made of 18 hexagonal- pieces. These pieces- need to line up perfectly- to create a smooth surface. The mirror is coated with gold to make it reflect infrared light better.

The telescope also has a sun shield that's as big as a tennis- court. This shield protects- the telescope from the sun, earth, and moon. It needs- to open up correctly after it's launched and stay in the right shape and position throughout the mission. The telescope also has four scientific tools that need to work in very cold temperatures.

It needs special cooling systems and heaters to keep them working. The telescope also has four scientific tools that need to work in very cold temperatures. It needs special cooling systems and heaters to keep them working. If any of these parts break- or get damaged, the telescope won't work- well or could even become- useless.

Launch- and deployment risks: The JWST- is too big to fit- in any rocket, so it has to be folded- and stored for launch.

When it gets- to its orbit, about 1.5 million kilometers- from Earth, a complicated and dangerous- process is needed- to unfold it. This process has over 300 steps. For example, you have to uncover, unlatch, and unhinge. Then, you need to extend, connect, and activate different parts. It will take about a month to finish. During that time, the telescope will face extreme temperatures, vibrations, shocks, and tiny space rocks. If anything goes wrong during this time, the telescope might not work right or get damaged. Most of the important systems on the telescope don't have backups. There's no way to fix them if something bad happens.

Cost and schedule- overruns: The JWST is a very expensive and complex- science project by NASA. It was supposed to cost- about $1 billion, but it ended up costing about $9.8 billion. It was supposed- to launch in 2007. However, it has been delayed for over 16 years due to technical challenges, design changes, testing issues, management problems, and budget cuts. These- delays caused a lot of problems for NASA's budget and resources. They also affected the scientific community and the public's expectations. It also reduced the money and chances for other astronomy missions and programs. The JWST is an amazing scientific project. It will help us learn more about how the universe started and changed over time. But it also has big problems and dangers that could stop it from working or even getting to where it needs to go. When NASA launches and sends out the JWST, it will be a really important time. After many years of hard work, we hope it will be successful.

CHAPTER THREE
THE JAMES WEBB TELESCOPE SNAPS STUNNING SHOT OF URANUS

The James Webb Space Telescope (JWST), an exceptionally powerful space observatory, has unveiled a fresh depiction of Uranus and its icy rings. The European Space Agency (ESA) disclosed this image on April 6, 2022. This portrayal offers a remarkably detailed view of 11 of the 13 recognized rings encircling the seventh planet from the Sun.

JWST revealed a new image of Uranus- and its icy rings. The European Space Agency (ESA) shared- this image on April 6, 2022. The image shows- 11 of the 13 rings around Uranus in great- detail. Some rings are so close- together that they appear to merge into one larger ring. The ESA highlighted this in a tweet. JWST used its infrared sensors to capture this image. This allows it to capture more light from faraway objects.

According to the ESA, the image reveals intricate details of Uranus's atmosphere. It also shows how it changes over time. The faintest rings of Uranus were last seen by the ground-based Keck Observatory in Hawaii. The Voyager 2 spacecraft also saw them in 1985. Voyager's images showed Uranus as a blue sphere. This is because of its dense methane gas atmosphere.

The James Webb Space Telescope / Brian S.A. Corey

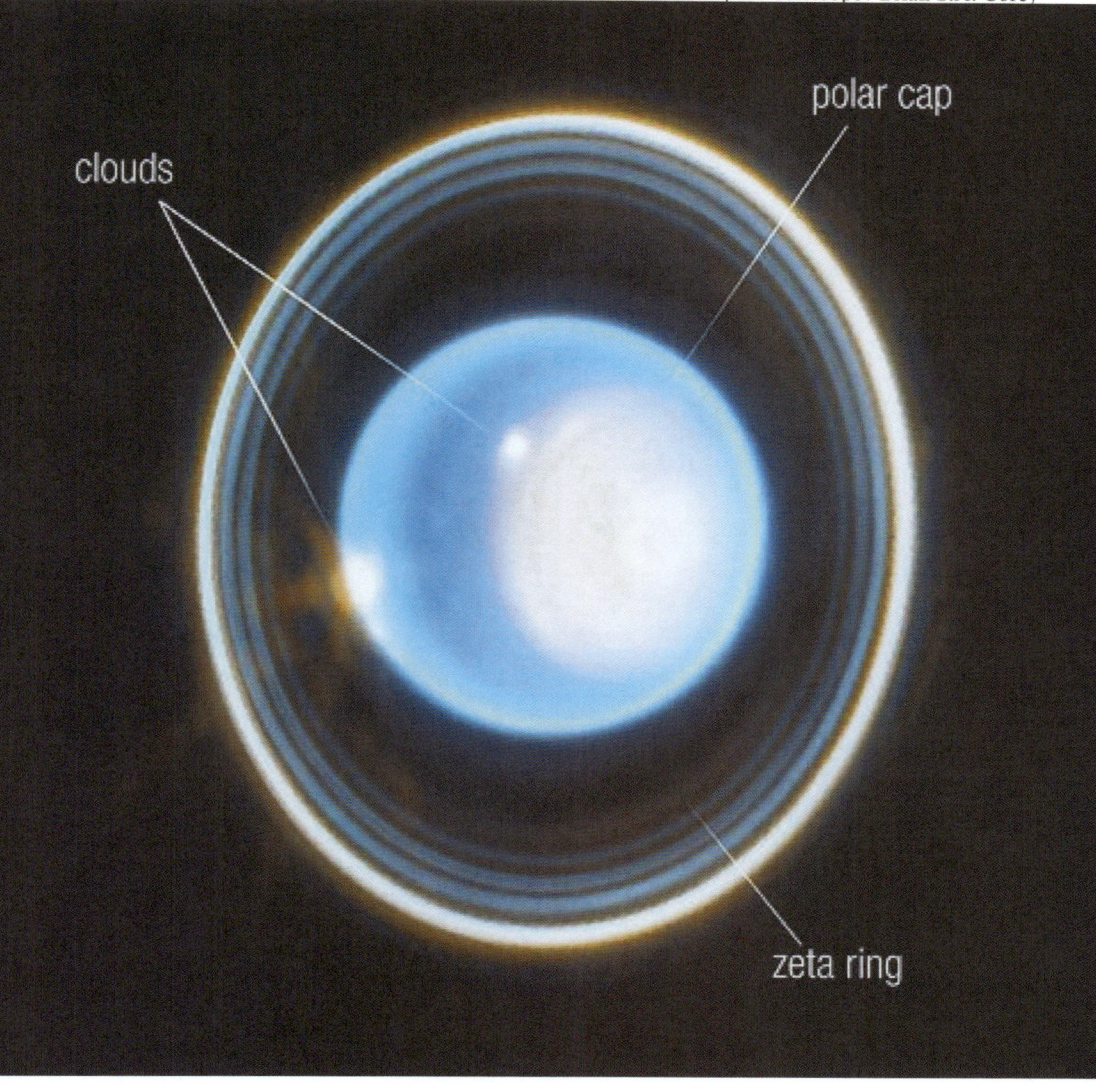

clouds

polar cap

zeta ring

The JWST also found changes- in Uranus's atmosphere. There is a bright area on the right- side of the planet, called a polar cap, which is visible- when the pole faces the Sun. This polar cap is unique to Uranus and gets brighter in the summer but fades away- in the autumn.

The space telescope also discovered a brighter spot in the center of the polar cap. Previous telescopes never saw it. The image- also shows two bright clouds, one along the edge of the polar cap and another on the planet's edge. The ESA thinks that both of these clouds are probably connected to storms.

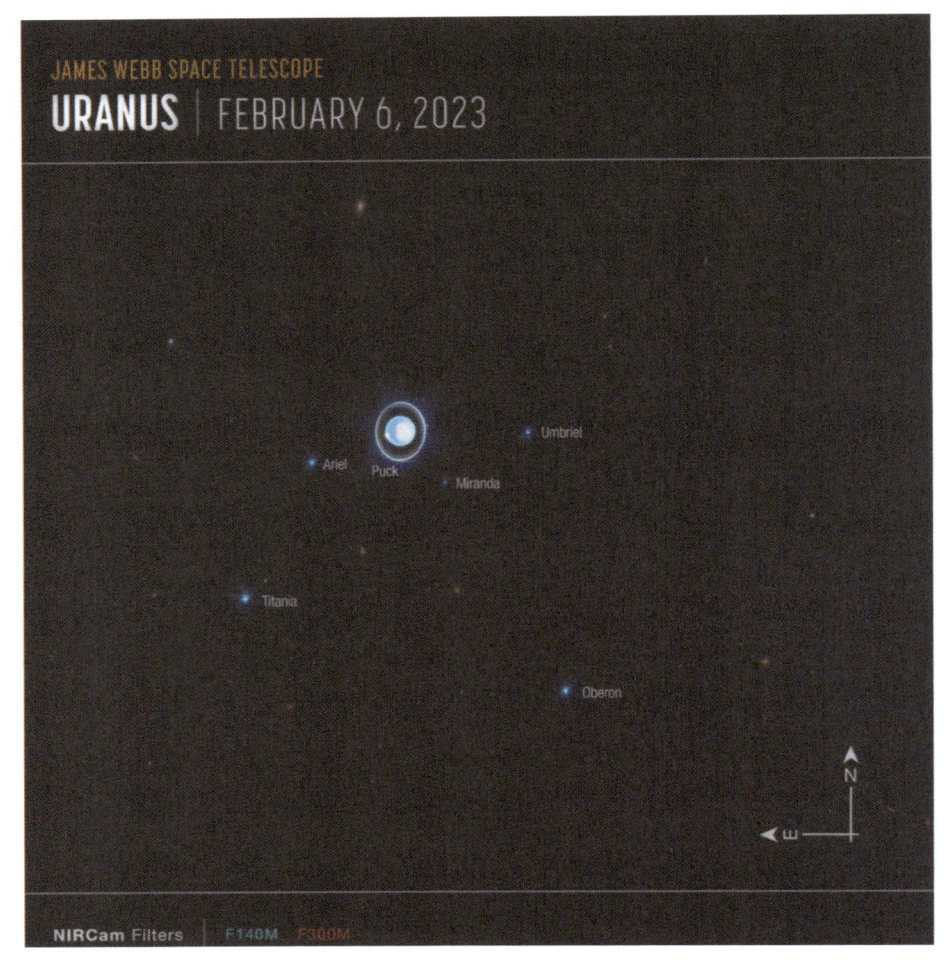

While observing, JWST captured Uranus and six of its moons. Uranus- has 27 moons, but most are too small and faint- to see.

A phenomenon New James Webb telescope photo of a galaxy cluster 6 billion light years

NASA's James Webb telescope released new images of galactic "arcs and streaks" in space. The images result from a cosmic phenomenon called gravitational lensing.

Gravitational lensing happens when a big celestial body distorts spacetime. It bends light around it, making distant galaxies look bigger. The European Space Agency says this happens when a celestial object with strong gravity bends the path of light, like a lens. This is because spacetime is curved.

The Phantom Galaxy, also called- M74, is shown in this picture. It is located- 32 million light-years away and has about- 100 billion stars. The picture combines images- from the James Webb Space Telescope- and the Hubble Space- Telescope.

NASA researchers used the JWT to take pictures of galaxies. They used infrared light. They wanted to learn more about how stars, gas, and dust form in nearby galaxies. These galaxies hadn't been seen before. The researchers published the pictures in a press release.

According to NASA's statement, we collected data and ended up with 21 research papers. These papers explain how small processes in our universe, like the early stages of star formation, affect the development of big things, such as galaxies, in our cosmos.

The main picture shows the bright center of a spiral galaxy. The Webb telescope allows us to see this. The galaxy's infrared light looks gray. The white color shows that there is a lot of star formation happening at the center.

The Webb telescope team is studying 19 spiral galaxies. This gives them a very detailed view. The telescope lets scientists watch how the energy from young stars affects the gas nearby. NASA says they have already started observing five of the 19 galaxies with the Webb telescope.

David Thilker from Johns Hopkins University in Baltimore, Maryland, was surprised. He said, "We were caught off guard by how clearly we could see the intricate formations."

Erik Rosolowsky, a researcher at the University of Alberta, said, "We are seeing how energy from new stars affects the surrounding gas, and it's amazing."

The next image shows a spiral galaxy with a web-like look and a very bright center, as seen by the Webb telescope. The galaxy's infrared light is shown in gray, and the red spikes come from the telescope diffraction. The colored dots in the background are other galaxies.

The Webb telescope helps astronomers see clear pictures of how stars form. Before, these early stages were hidden. NASA reports that Adam Leroy, a researcher at Ohio State University, said the process of star formation was covered by gas and dust clouds. It couldn't be seen until now.

The Webb telescope can see further than ever before because its infrared vision can penetrate dust.

NASA describes the third galaxy image as a close-up of a wispy spiral galaxy. The image also shows tiny red dots, which are other galaxies in the background.

Gravitational lensing is when spacetime gets warped by a huge celestial body. This makes the light from galaxies and stars behind it bend. Distant observers see a messed up view because of this.

Scientists use gravitational lensing to magnify distant galaxies for better study. The SDSS J1226+2149 galaxy cluster is in the Coma Berenices constellation. It is about 6.3 billion light-years away, according to the ESA.

NIRCam is Webb's main camera for near-infrared images. It captured- a clearer and brighter image of the Cosmic- Seahorse galaxy. The galaxy appears- as a long, bright, and distorted arc near the core- in the lower right of the image.

The James Webb Space Telescope / Brian S.A. Corey

NASA's Webb Telescope Captures Rarely-Seen Prelude to Supernova

The bright and hot star Wolf-Rayet 124 (WR 124) stands out at the heart of the James Webb Space Telescope's composite image. The image combines near-infrared and mid-infrared light from two of Webb's instruments: the Near-Infrared Camera and the Mid-Infrared Instrument.

A dazzling star- dominates the scene, surrounded by smaller twinkles of light. A ring- of gas and dust- encircles the star, thicker- at the top and bottom than on the sides. This is WR 124, a rare and massive star that NASA's James Webb Space Telescope captured- in June 2022. Webb's infrared capabilities- unveil the star and its nebula with breathtaking- precision. Positioned- 15,000 light-years away within the Sagitta- constellation, the star in focus- is WR 124.

WR 124 is a large- and bright star- that doesn't live for very long. It's a specific- kind of star called a Wolf-Rayet star, which means it's getting- close to the end- of its life. These stars quickly lose their outer layers, creating amazing rings of gas and particles. WR 124 is 30 times heavier than the Sun and has already let out the same amount of stuff as ten suns. When the gas- cools down, it turns into space dust and gives off infrared light.

Cosmic dust is crucial for- the universe. It helps create new stars and planets and provides a safe place for complex molecules to develop. Cosmic dust might even be where life on Earth comes from. But astronomers still need to figure out how cosmic dust- is formed and how it survives a supernova explosion. There's more dust in the universe than current theories can account for.

Webb's instruments can- solve this mystery. Webb's NIRCam can balance the brightness of WR 124's core and the fainter details in its nebula. Webb's MIRI can show the clumpy structure of the gas and dust around the star. Astronomers can use Webb's data to study- the amount of dust produced by stars like WR 124 and the size and durability of the dust grains.

Stars like WR 124 show us the early history- of the universe. These stars were the first to make heavy elements- in their cores, which are now common on our planet.

Webb captured- a picture of WR 124 during a short and chaotic period of star development. This picture could lead to new discoveries- about space dust.

NASA's Webb Spots Swirling, Gritty Clouds on a Remote Planet

The James Webb Space Telescope by NASA has unveiled the presence of silicate clouds within the atmosphere of a remote planet. The planet, called VHS 1256 b, has a 22-hour day and a very dynamic atmosphere. The hot air rises, and the cold air sinks, creating huge variations in brightness. It is the most variable object of its size ever observed.

The research- team, led by Brittany Miles from the University of Arizona, also detected water, methane, carbon monoxide, and carbon dioxide in the planet's-atmosphere. This is the most molecules ever found at the same- time on an exoplanet.

VHS 1256 b is a unique planet. It circles two stars, not one, for 10,000 years. It is also far from its stars, even farther than Pluto is from our sun. Miles said, "This makes VHS 1256 b a great target for Webb. The planet's light isn't mixed with the stars' light." The planet's atmosphere is hot at high altitudes. There, silicate clouds swirl. The temperature- can reach 1,500 degrees Fahrenheit- (830 degrees Celsius).

The James Webb Space Telescope / Brian S.A. Corey

NASA's JWST- has discovered turbulent clouds in the atmosphere of VHS 1256 b, a distant- planet. This planet orbits around two rapidly rotating stars. The clouds in its sky are made of silicate dust. They constantly change throughout its 22-hour day.

The planet's analysis shows- clouds made of different-sized silicate dust particles. Beth Biller, a co-author from the University of Edinburgh in Scotland, explains that some of these particles look like tiny smoke particles. Others look like very small, very hot grains of sand. The Webb telescope can see these clouds because the planet is young and has low gravity. It is only 150 million years old. This makes the atmosphere more turbulent and full of clouds.

The research team was amazed- by the spectrum's complexity. They compared it to a treasure trove of data. They found silicate dust, but- they also wanted to learn about the clouds and their characteristics. Miles, a team member, said, "This is just- the beginning. We're starting a project to understand this planet and interpret the data from Webb."

The spectrum showed molecules that give us information about the planet's climate and weather. Other telescopes have observed similar things on other planets, but only Webb observed multiple things at once on one planet. Andrew Skemer, a scientist from the University of California, Santa Cruz, says, "With one spectrum from Webb, we can see numerous molecules. These molecules reveal the planet's clouds and weather."

EXOPLANET VHS 1256 b
EMISSION SPECTRUM

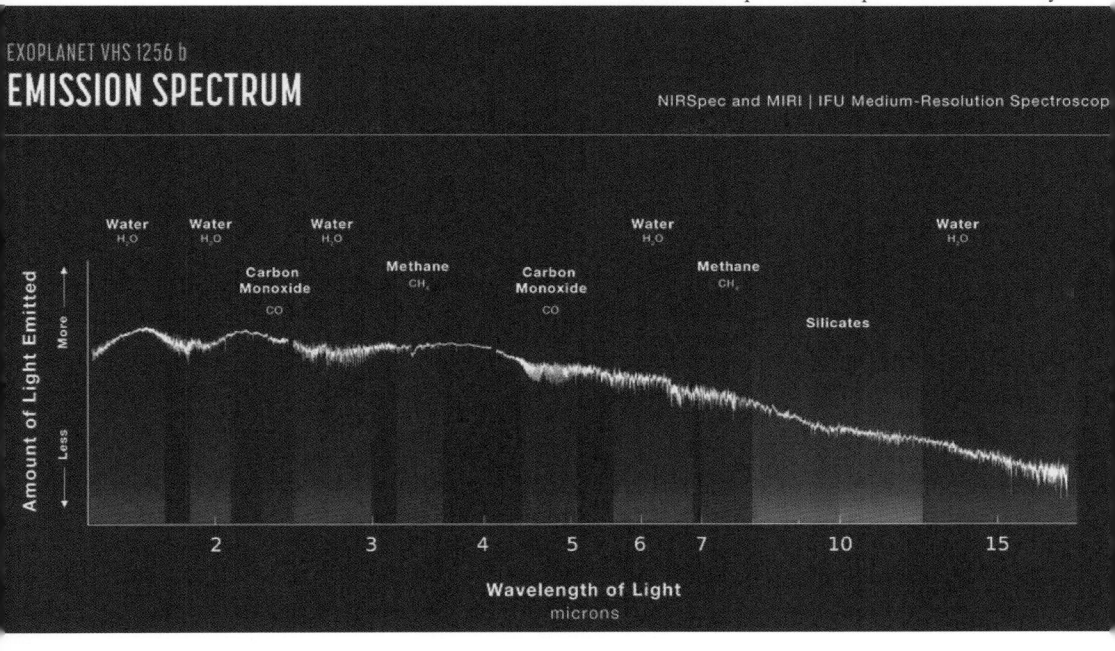

NIRSpec and MIRI | IFU Medium-Resolution Spectroscop

The James Webb- Space Telescope discovered the secrets of planet VHS 1256 b. It used two spectrographs to measure- the planet's infrared light. The light revealed silicate clouds in the planet's atmosphere. The light also revealed- water, methane, carbon monoxide, and carbon dioxide- in the planet's atmosphere. The researchers were able to see the planet directly because it orbits at a great distance from its stars. The data from Webb is detailed and rich. The researchers expect to make more discoveries about VHS 1256 b in the future. Biller remarked- that the payoff is substantial, even though the telescope usage- is relatively small. With just a few hours of observations, it feels like we've opened up endless possibilities for more discoveries. The planet is young and hot, only 150 million- years old.

It will cool as it gets older, and its clouds- may clear. The researchers are curious about its appearance in billions of years.

The researchers are part- of Webb's Early Release Science program. The program aims- to help astronomers learn how to use the telescope. They want to study planets and how they form.

CHAPTER FOUR
LAUNCH AND DEPLOYMENT

The JWST was launched and deployed in Kourou, French Guiana, using an Ariane 5 rocket on December 25, 2021.

When and how was it launched- into space?

The JWST- went into space- on December 25, 2021. It launched- at 12:20 UTC on an Ariane 5 rocket- from Europe's Spaceport- in French Guiana. NASA, ESA, and CSA worked together on this launch. They provided the rocket, launch site, and some scientific instruments for the telescope.

The launch process was complex and risky. It had several stages and maneuvers. The Ariane five rocket lifted off from the launch pad. It went up through the atmosphere. It reached a speed of about- 8.4 km/s (about 18,800 mph) and a height- of about- 150 km (about 93 miles). The rocket then separated- into two parts. The main stage fell back to Earth. The upper stage carried the Webb telescope into orbit.

The upper stage did two engine burns to go higher and reach a point to let go of the Webb telescope. The first burn was about 16 minutes long and made the upper stage go faster and higher. The second burn was about 3 minutes long and made the orbit a circle at about 1,400 km (about 870 miles) high.

The second burn was about 3 minutes long and made the orbit a circle at about 1,400 km (about 870 miles) high. After about 27 minutes from liftoff, the upper stage let go of the Webb telescope. Then the telescope opened up its solar panels and communication antenna.

The launch was successful. It marked the beginning of Webb's journey to its final destination: the second Lagrange point (L2). L2 is a stable location in space, about 1.5 million kilometers (930,000 miles) from Earth. Webb will revolve around the Sun there, synced with our planet. To get to L2, Webb had to make course corrections with its thrusters and unfold its complex parts. These parts include the primary mirror, secondary mirror, sun shield, and instruments. This process took 29 days and NASA called it "29 days on the edge".

The JWST is now at L2 and studying the universe's mysteries. It uses its strong infrared vision to study the first galaxies that formed after the big bang. It will also study the formation of stars and planets. Lastly, it will search for life outside of Earth. This is a new discovery for the Hubble Space Telescope. It has been observing the universe for over 30 years.

How did it unfold- and deploy its primary mirror- and sun shield?

The James Webb Space Telescope has a big- and complicated mirror and shield. They were- folded and stored inside the rocket- for launch. In space, the telescope had to do a series- of steps to unfold and deploy it. It took about 29 days- to finish. Here is a short summary of how it unfolded and deployed the mirror and shield:

The telescope was separated- from the rocket. It then activated its solar panel to obtain electricity. It also activated its antenna to communicate- with Earth. The telescope then employed its thrusters to correct its path. This was done- in order to get to its eventual destination, the second Lagrange point (L2). This location is in space, approximately 1.5 million kilometers from Earth. The telescope's secondary mirror support structure was then enlarged. Three booms on the structure hold the secondary mirror in front of the primary mirror.

The telescope unfolded its primary- mirror wings. These wings are two mirror segments hinged to the central segment. The wings were extended and latched into place. This formed a 6.5-meter hexagonal- mirror made of 18 individual segments.

The telescope deployed its sun shield. This shield is a five-layer membrane. It protects the telescope from the heat- and light of the sun, earth, and moon. The sun shield was folded like origami and stored on a pallet. The telescope released the pallet. Then, it deployed the sun shield layers one by one. This was done using cables, motors, and springs. The sun shield also deployed its tensioning system. This system- consists of masts, booms, and cables. These keep the layers taut and aligned.

The telescope- checked and calibrated its components to make sure they were ready for science- operations. The unfolding and deployment- of the primary mirror and sun shield was a critical and challenging- phase of the mission. If there had been- any failure or problem, the whole mission could have been in danger. However, the telescope and its parts were engineered- and tested carefully. The unfolding and deployment succeeded. This was a big accomplishment- for the James Webb Space Telescope.

How did it reach its final- orbit around- the Sun-Earth Lagrange point 2 (L2)?

The James- Webb Space Telescope (JWST) completed- its final orbit- around the Sun-Earth Lagrange point 2 (L2). After separating- from the rocket's upper stage, it used its thrusters- to make multiple course adjustments. L2 is a location in space- where the Sun's and Earth's gravitational pulls balance- each other. This permits a spacecraft to circle both- the sun and the earth. L2 is around 1.5 million kilometers distant from Earth, facing away from the Sun.

The telescope had to follow a complex path to reach L2. It started- its journey on December 25, 2021, at 12:20 UTC3. An Ariane 5 rocket- from Europe's- Spaceport in French Guiana launched- the telescope. The rocket- went up into the sky and reached- a speed- of about 8.4 km/s and an altitude- of around 150 km. After- that, the rocket split into two parts. The main stage fell back to Earth, and the upper stage carried the Webb telescope into orbit.

The upper stage did two engine burns to raise its orbit and reach a point where it could release the Webb telescope. The first burn lasted about 16 minutes and made the upper stage go faster and higher. The second burn lasted about 3 minutes and made the orbit circular at an altitude of about

1,400 km. After about 27 minutes from liftoff, the upper stage released the Webb telescope. Then, the telescope deployed its solar array and communication antenna.

The telescope used its thrusters to make course corrections and reach L2. The first correction happened 12 hours after the launch. Three more corrections followed over the next two days. The last correction happened 29 days after launch, when the telescope reached L24. After that, the telescope entered a halo orbit around L2. This orbit takes about six months to complete.

Getting to L2 was a crucial and tough part of the mission. Any mistake could have stopped the telescope from getting to the best spot for science work. But, thanks to careful work on the telescope and its parts, the trip to L2 was successful. This is a big- accomplishment for the James Webb Space- Telescope.

How does it observe and collect- data from the infrared universe?

The James Webb Space Telescope- (JWST) is an amazing tool. It can observe- and collect data from the infrared universe. Infrared- light is a type of light that has longer wavelengths- than visible light. It can't be seen- by humans. But, infrared light can uncover secrets of the cosmos. These secrets are hidden by dust, gas, and distance. Let me explain how JWST does this:

The JWST has a big mirror. The mirror has 18 hexagonal segments. The segments make a 6.5-meter wide surface. This mirror- can gather more light than any other space telescope- before it. It can focus the light onto four scientific instruments. The instruments- are located behind the mirror.

The JWST has a sun shield- with five layers. This shield is designed to protect the telescope from the sun, moon, and earth's heat- and light. It keeps the telescope at the right temperature- to detect faraway objects' infrared radiation. It also stops unwanted light from reaching the instruments. MIRI is a tool that- can study images and spectra- in the mid-infrared range, which- is from 5 to 28 micrometres. It can block the light from bright stars using a device- called a coronagraph.

The JWST has four scientific instruments. These instruments can observe infrared light at different wavelengths and resolutions. The instruments are the Near-Infrared Camera (NIRCam), the Near-Infrared Spectrograph (NIRSpec), the Mid-Infrared Instrument (MIRI), the Slitless Spectrograph (FGS/NIRISS), and the Fine Guidance Sensor/Near-Infrared Imager.

NIRCam is a camera. It can take pictures of the sky. It uses near-infrared light. The range is from 0.6 to 5 micrometers. NIRCam can split the light. It can come in different colors. It does this using filters and gratings. It can also measure the spectra of objects. NIRCam can see far-away galaxies. It can see galaxies from the early universe. It can also study star and planet formation in our galaxy.

NIRSpec is a powerful tool. It can analyze light from many objects at the same time. It works in the near-infrared range, from 0.6 to 5 micrometres. NIRSpec can also choose which objects to observe using a special micro shutter array. It can measure the chemical makeup, temperature, movement, and distance of stars, galaxies, and other celestial bodies. This allows us to see fainter objects, like planets or dust disks, that are close to the stars. MIRI can also explore warm objects that give off infrared radiation, such as young stars, protoplanetary disks, exoplanets, asteroids, comets, and brown dwarfs.

FGS/NIRISS is an instrument with two functions. It guides and stabilizes the telescope using a fine guidance sensor (FGS). It also observes- images and spectra in the near-infrared range of 0.8 to 5 micrometres- using- a near-infrared imager and slitless spectrograph (NIRISS). FGS/NIRISS can resolve- small details on the surfaces of stars using aperture masking interferometry. It can also detect- the atmospheres of exoplanets- using transit spectroscopy.

What are the main scientific instruments and modes of observation?

The JWST is a telescope that will study the universe using infrared light. It will use four scientific instruments to do this. One of these instruments is called the Near-Infrared Camera (NIRCam). It will take high-resolution pictures of the sky using infrared light. NIRCam will also have special filters to block out bright light from stars. This will help scientists see faint objects like planets or disks around stars. NIRCam will be used to study stars, galaxies, planets, and exoplanets.

The Near-Infrared Spectrograph (NIRSpec) is an instrument. It will measure the spectra of infrared light from stars, galaxies, and nebulae. A spectrum is a plot of light intensity at different wavelengths. It can tell us about the source's composition, temperature, motion, and distance. NIRSpec has four modes of operation: multi-object mode, fixed-slit mode, integral field unit mode, and prism mode. In multi-object mode, it can observe up to 100 sources at once. In fixed-slit mode, it can observe one or a few sources at a time. In integral field unit mode, it can obtain spectra over a small area of the sky. In prism mode, it can obtain low-resolution spectra over a wide range of wavelengths. NIRSpec will study the formation and evolution of galaxies.

It will also study the chemical enrichment of the interstellar medium. Additionally, it will examine the physical conditions and movement of gas in galactic nuclei. NIRSpec will study the atmospheres of exoplanets.

The Mid-Infrared Instrument (MIRI) will observe the infrared sky in two wavelength bands: 5–12 microns and 12-28 microns. MIRI will have a coronagraph and an integral field unit mode. It can detect faint and cold objects better than NIRCam and NIRSpec. These objects include brown dwarfs, young stellar objects, protoplanetary disks, and distant galaxies. MIRI can also study the dust and ice features in the spectra of these objects. This can give us clues about how they formed and evolved. To operate at a temperature of about 7 K (-266 °C), MIRI will need active cooling. A cryocooler will provide this cooling.

The FGS/NIRISS performs two key roles. It directs the telescope by measuring the location and velocity of guide stars. Second, it conducts scientific observations in the near-infrared spectrum of 0.8 to 5 microns. The instrument supports four observation modes:

1. Wide-field imaging: This mode captures- images over a large area of the sky.
2. Single-object slitless spectroscopy- obtains spectra of bright sources- without a slit.
3. Aperture-masking interferometry- enhances image resolution. It uses a mask with holes in the telescope aperture.
4. Non-redundant aperture- masking obtains high-contrast images of faint sources near bright ones.

The JWST is being fixed and upgraded by a human orbital maintenance platform (OMP). In this photo, you can see the front of the JWST. The OMP is lighting up the gold-coated main mirror.

What are some of the expected discoveries and benefits of the mission?

The JWST is an amazing mission. It changes how we understand the universe. The telescope- looks at the universe using infrared- light. This lets it see things that optical- telescopes, like the Hubble Space Telescope, can't see. The mission will make many important discoveries and bring many benefits.

The first light and reionization:

The JWST will look back in time. It will see the first stars and galaxies. They formed after the Big Bang. The universe was only a few hundred million years old. It will also study how these objects ionized the intergalactic medium. This made it transparent to light. This will- help us learn about the origin- and evolution of cosmic structures. It will also help- us learn about the role of dark matter and dark energy- in shaping them.

The assembly- of galaxies:

The JWST will study how galaxies- formed and changed over- time. It will measure things like galaxy size, shape, and mass. It will also measure star metal content, and black hole activity.

It will also show how galaxies collide, exchange energy, and get distorted by gravity.

The first light and reionization:

The JWST will look back in time. It will see the first stars and galaxies. They formed after the Big Bang. The universe was only a few hundred million years old. It will also study how these objects ionized the intergalactic medium. This made it transparent to light. This will- help us learn about the origin- and evolution of cosmic structures. It will also help- us learn about the role of dark matter and dark energy- in shaping them.

The birth of stars & protoplanetary systems:

The JWST will look at dusty regions where stars and planets are made. Optical telescopes can't see these regions. The JWST will- see individual stars and their- disks, jets, and outflows. It will also study the chemistry- and physics of the stuff between- the stars. This stuff makes stars and planets. The JWST will study how different types of stars affect their surroundings. It will also study how they affect planets' ability to support life.

The beginnings- of planetary systems- and life:

The JWST is a telescope- that will study many planetary systems, including- the ones in our solar system. It will look- at planets, moons, asteroids, comets, and objects- in the Kuiper belt. This will help us understand- their geology, climate, atmosphere, and potential- for life. It will also find- and study exoplanets around other- stars. It will use different techniques- like transit spectroscopy, direct imaging, and microlensing. The JWST will learn about- their orbits, masses, sizes, temperatures, compositions, and atmospheres. It will also look for signs of life- in their atmospheres, like water, oxygen, methane, or other things that show- there might be life.

CHAPTER FIVE

FIRST PICTURES OF MARS TAKEN BY THE JAMES WEBB SPACE TELESCOPE

NASA launched the (JWST) on December 25, 2021. It is the largest telescope in space. It can see things that the Hubble Telescope cannot because they are too old, distant, or faint. The JWST has high resolution and sensitivity.

The James Webb Space Telescope captured its first pictures and spectra of Mars. On Sept. 5, it used its Near-Infrared Camera (NIRCam) to take images of the eastern side of the planet. The camera captured the planet in different infrared colors.

The first picture is a combination of two images. It shows a map of Mars' surface created by NASA and the Mars Orbiter Laser Altimeter (MOLA). The areas covered by the NIRCam instrument are marked on the map.

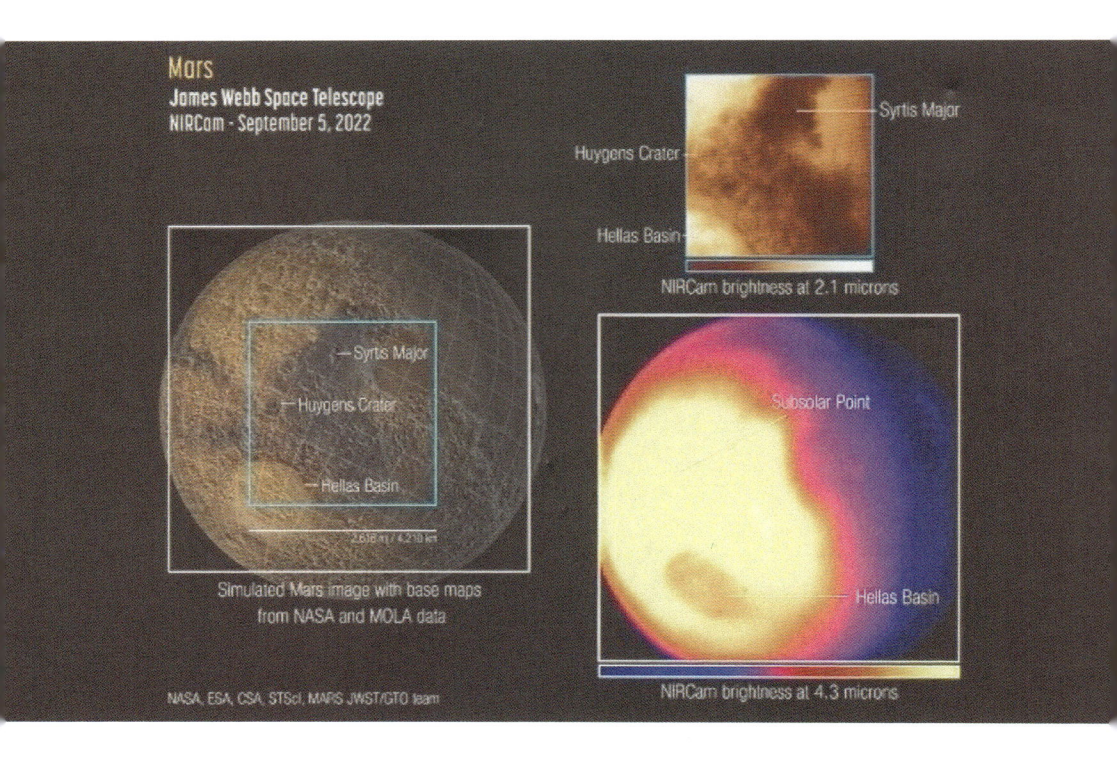

74

The image above, top right, shows Mars reflecting sunlight captured- using shorter wavelengths. This image- reveals distinct features. These include the rings of the Huygens- Crater, the dark lava plains of Syrtis Major, and the brightening of the Hellas Basin. On the other hand, the image- above, bottom right, shows Mars emitting light as it cools down. The brightness- is highest when the Sun is directly overhead and decreases towards the poles. However, some light is absorbed- by carbon dioxide molecules. This causes the Hellas Basin to appear darker than its surroundings. The Hellas- Basin is the largest and best-preserved impact crater on Mars. Geronimo Villanueva, the principal investigator at NASA's Goddard Space Flight Center and the architect- of these Webb observations, explains, "The appearance of Hellas is not caused by temperature changes. It will be interesting to study and understand these different effects in the data."

JWST is positioned almost 1 million miles away. It can capture the visible disk of Mars, which- is the side facing the telescope that is lit by the sun. Scientists can analyze the snapshots to study- dust storms, weather shifts, seasonal variations, and daily processes. However, there's a challenge. Mars- is close to Earth, so it shines very- brightly in the night sky.

This is good for people- who like to look at the stars, but not so good for an observatory that needs to see very faint light from faraway- galaxies. Astronomers have used special techniques to deal with Mars' brightness. This allows Webb to record small bursts of light without being overwhelmed. The team will use the pictures and data from Mars to learn about the differences in geography across the planet. They will also look for certain gases, like methane and hydrogen chloride, in Mars' atmosphere.

The James Webb Telescope- took a spooky picture of the Pillars- of Creation.

NASA- is well known for its depiction- of the "Pillars of Creation." These are an extensive dust cloud located light years away. They are where new stars are born. However, this new version gives a completely different perspective.

NASA published a new photograph of the Creation Pillars. The image was captured by the JWST . The photo has an unusual sense about it. It represents the Eagle Nebula's Pillars of Creation as a ghostly hand.

Warm gold and brown tones dominated Hubble Space Telescope images from 1995. They are now icy colors of chilling blue. It seems to be something from a terrifying crypt.

> NASA scientists have discovered a region where thousands of stars appear to disappear. This is because they are hidden by the mid-infrared light, which they do not emit, unlike the gas and dust.

The James Webb Space Telescope used- its mid-infrared sensors to take a photo. These sensors can see light- that human eyes cannot. They are good at showing- cosmic gas in detail. The Hubble- took an old photo of the Pillars of Creation using visible light, which is the light we can see.

The new picture from the James Webb Space Telescope- is on the right. It is next to the older photo from the Hubble Space- Telescope. (NASA)

NASA reported- that an interstellar dust covers the scene. The stars- are too faint to be visible in the mid-infrared light, which reveals the dust's location. The gas and dust pillars appear dull in color, but shine at their edges. This indicates- activity within them. NASA also mentioned that the red balls near the edges of the pillars are the young stars. These stars are obscured- by dust, but the mid-infrared sensors can detect them. NASA added that the dust is thickest where it appears grayest.

The James Webb Space Telescope / Brian S.A. Corey

The top of the scene resembles a red V-shape, where the dust is thinner and cooler, resembling an owl in flight.

The James Webb telescope used- its near-infrared sensor to take- a picture of the dust clouds. They looked different. The sensor showed many young stars forming- around the cosmic gas. The dust cloud- looks like a hand. It is much bigger- than our solar system, about four to five light-years long. The whole Eagle Nebula is even bigger, about 70 by 55 light-years in size. The nebula- is very far from Earth, about 7,000 light-years away.

The James Webb Space Telescope Snaps Image of Two Galaxies Merging

Two galaxies that- are more than 270 million light-years away- combine to form one galaxy. The JWST captured an image that shows this happening. The image- shows the two galaxies merging- and is called "IC 1623." According to the ESA, this merger- causes a lot of new stars to form. The gas in the galaxies is building up and creating stars very quickly. In fact, IC 1623 is making stars 20 times faster- than our Milky Way galaxy.

Astronomers took a picture by combining mid-infrared and near-infrared images of IC 1623.

The ESA says the starburst is very strong and emits a lot of infrared light. The merging galaxies might be creating a massive black hole. The James- Webb Space Telescope has more advanced- infrared sensors than- the Hubble- Space Telescope. Therefore, it can detect more stars forming in the galaxies' dust clouds.

The James Webb Space Telescope / Brian S.A. Corey

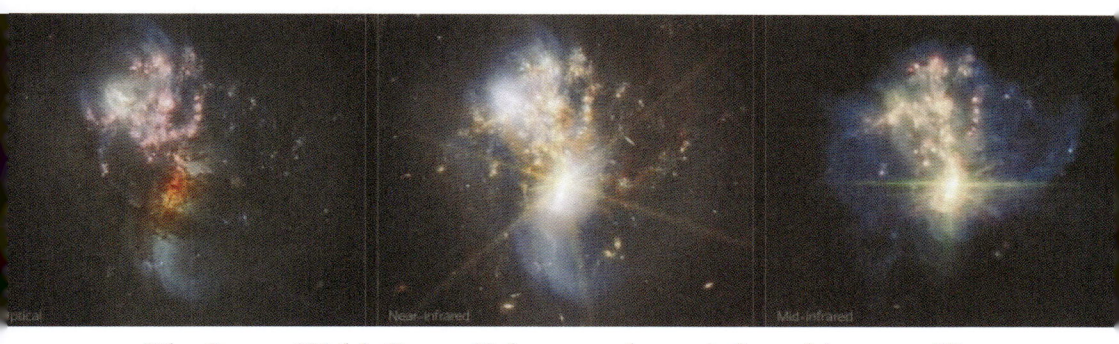

The James Webb Space Telescope shows infrared images. The Hubble shows pictures.

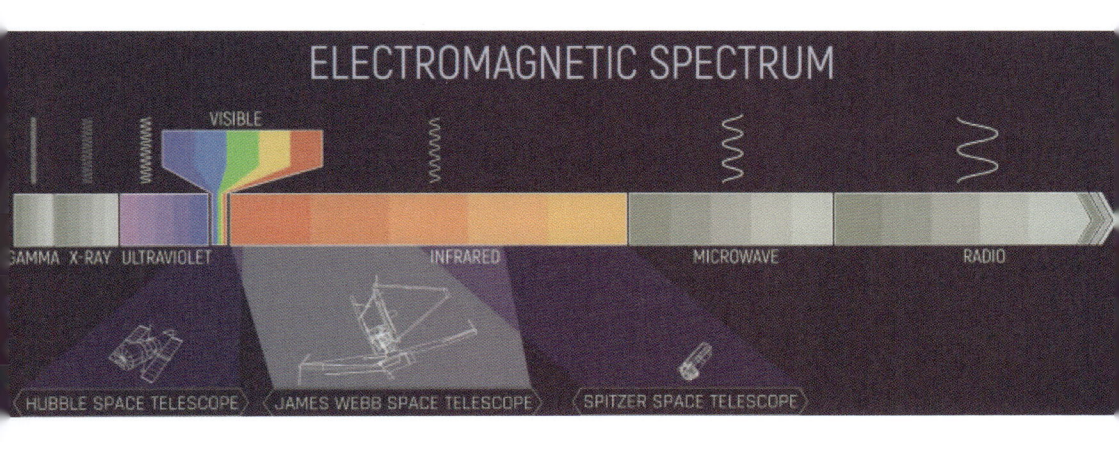

The ESA showed that the pictures of IC 1623 look different when taken with Hubble and James Webb. The new telescope can see more stars being made. It's especially good at spotting where two galaxies are colliding.

James Webb Telescope- Snaps Clearest Image of Neptune's Rings

The James Webb Space Telescope from- NASA shows- us Neptune's outer- rings. It took a picture- of this faraway planet, which- also caught its hard-to-see rings. NASA says the picture- is the best view of the planet's ring in thirty years. In 1988, the Voyager- space probe flew- by Neptune. It took photos of the rings- and the planet's blue atmosphere.

The James Webb telescope took a picture using strong infrared sensors. These sensors can see more light coming from Neptune and its icy rings than regular telescopes on the ground. Heidi Hammel, a scientist at James Webb, says this picture shows the faint and dusty rings that we haven't seen for thirty years. It's the first time we've seen them using infrared light. NASA also says the picture shows Neptune's less visible dust bands. It also shows some of its moons, including Galatea, Proteus, and Naiad.

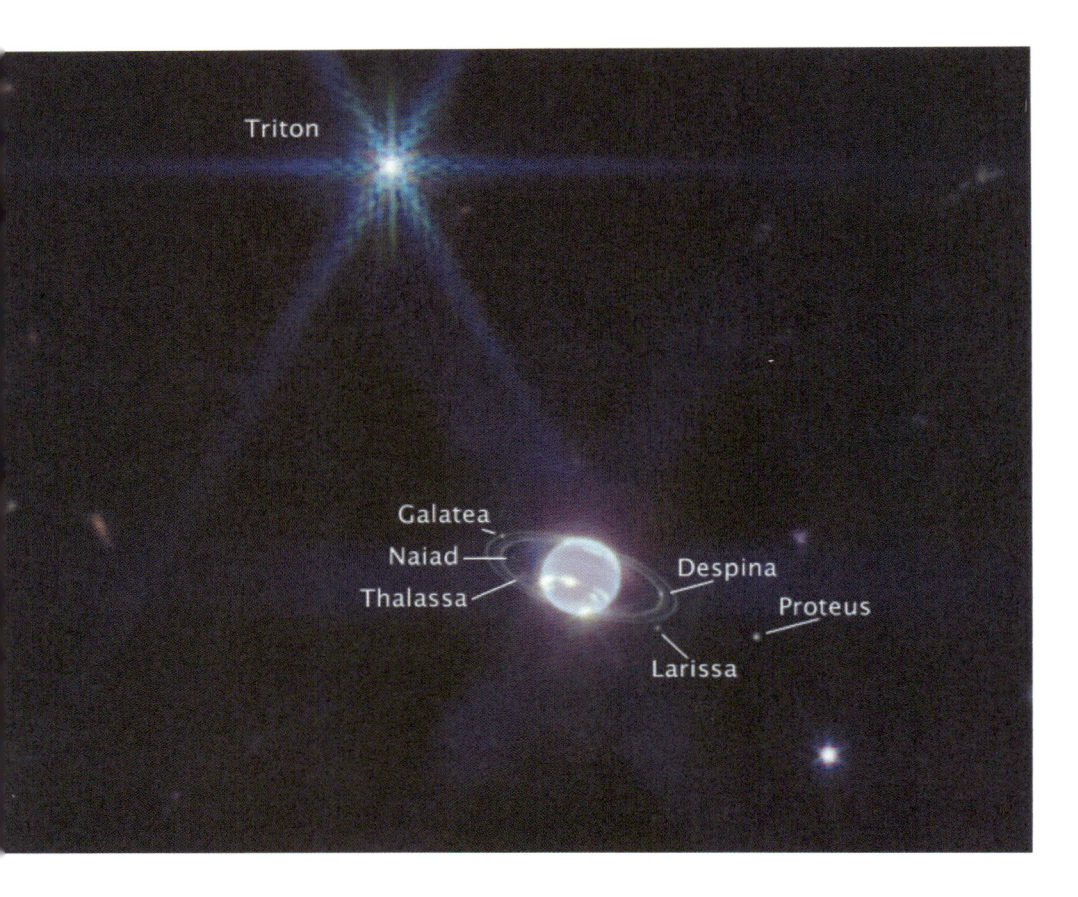

Seven of Neptune's fourteen known moons were seen by the JWST. These moons are named after the Roman God of the Sea. Triton, one of these moons, may be seen as a brilliant point of light in the upper left corner of the image. Triton circles Neptune in the opposite direction that it rotates. The telescope discovered methane absorption wavelengths, which darken the planet's atmosphere. As a result, it seems brighter than Neptune.

The James Webb image shows Neptune differently because it was taken in infrared. The Voyager pictures showed Neptune as blue. NASA explains that methane gas in Neptune's atmosphere makes the planet appear very dark in near-infrared wavelengths.

This is true except where there are high clouds. These methane-ice clouds look bright as streaks and spots because they reflect sunlight before the methane gas absorbs it. There is also a thin line of brightness around Neptune's equator. This may- indicate how the air moves and causes Neptune's winds- and storms. In the upper-left corner of the image, there- is a star, but it is actually- Triton, Neptune's largest moon. Triton has a frozen- layer of nitrogen- that reflects 70% of the sunlight it receives. In this image, Triton appears- much brighter than Neptune. The planet's atmosphere appears darker because of methane absorption in the near-infrared wavelengths.

NGC 604: This nebula has clumpy, red filamentary clouds. In the middle, there is a big bubble with an opaque bluish light and stars scattered throughout. The dust around the bubble seems white, and there are a few smaller bubbles around. Thousands of stars, generally yellow or white, populate the space around the nebula.

88

The James Webb Space Telescope / Brian S.A. Corey

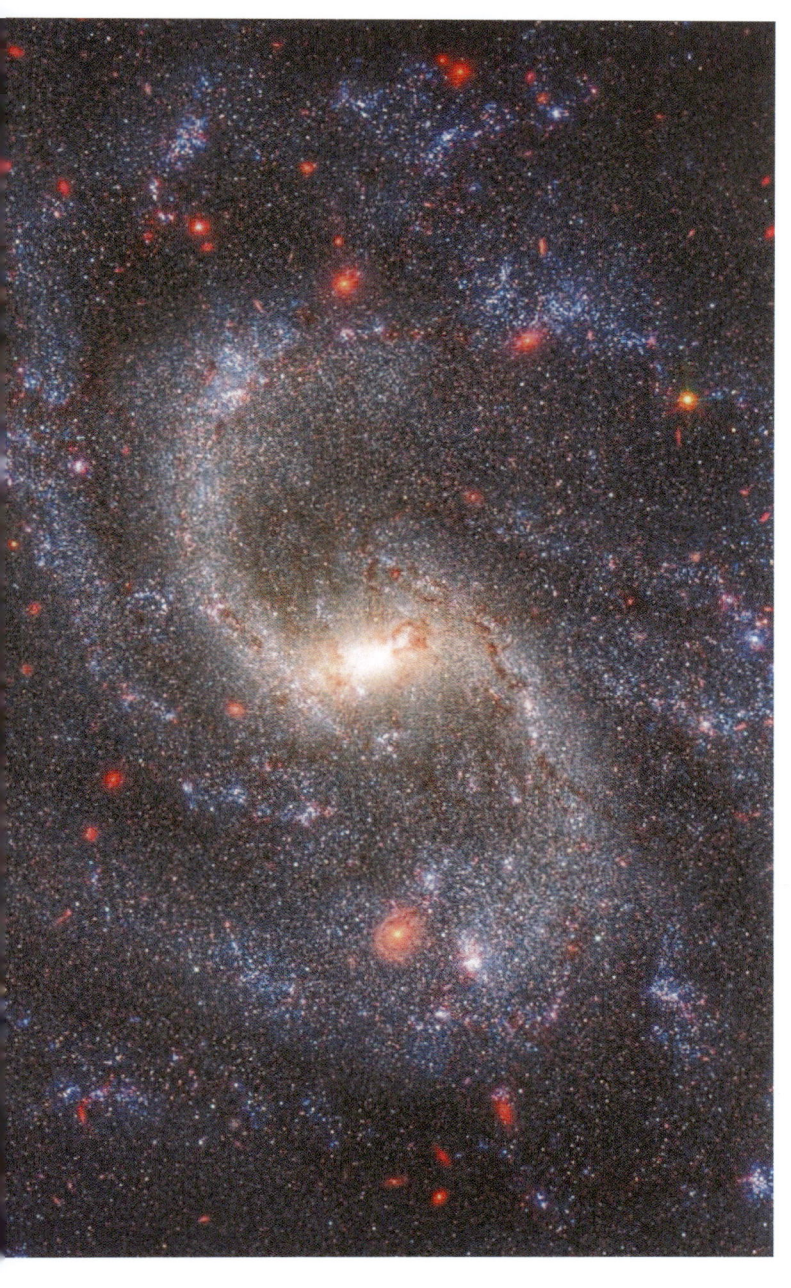

NGC 5468 is a face-on spiral galaxy with four spiral arms extending outward. The arms are loaded with young blue stars and speckled with purple star-forming areas. In the center, there is a brilliant, yellowish core with a narrow linear bar. Red background galaxies are spread across the photo. The photo has a dark space background.

This is another nebula. It has wispy, light blue clouds and a big, cavernous bubble in the center-right. Pink and white hues cover the bottom left side of the bubble, with hundreds of faint stars encircling it.

The James Webb Space Telescope / Brian S.A. Corey

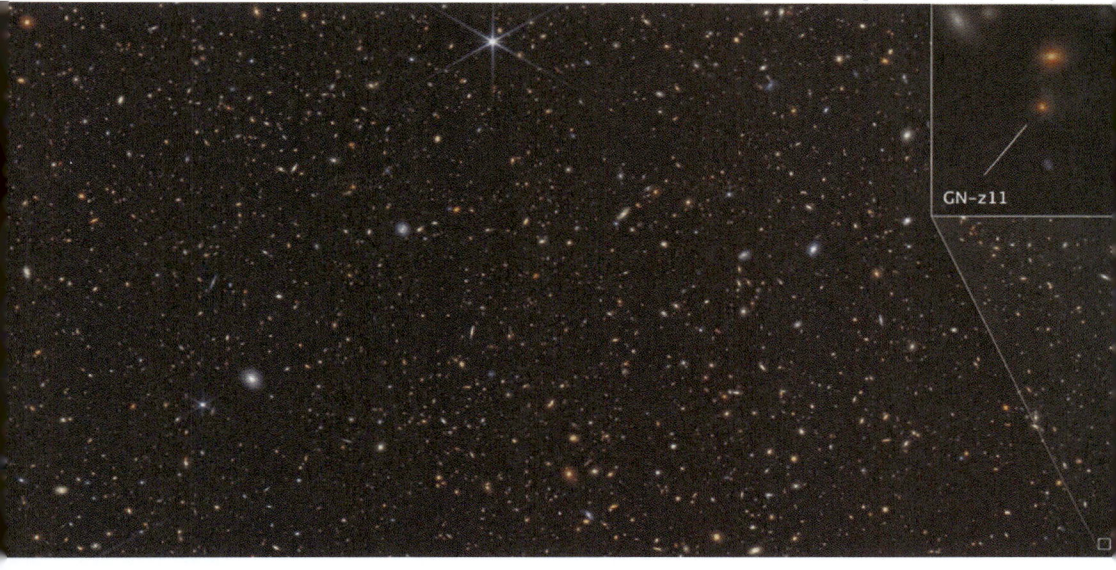

This rectangle picture depicts hundreds of galaxies. They come in all forms and hues. They are set on a black space background. Some galaxies are spirals, while others have blobby elliptical shapes. One conspicuous foreground star has diffraction spikes. A highlighted region focuses on a galaxy called GN-z11, which appears as a hazy yellow dot.

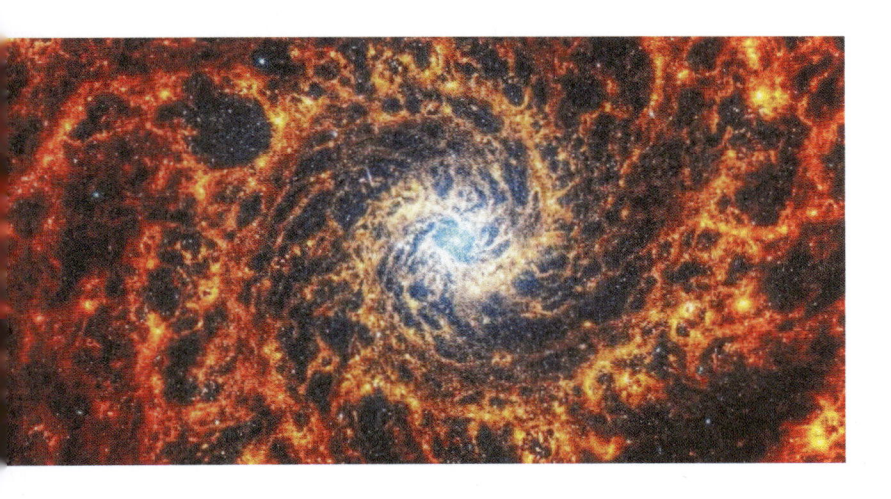

NGC 628 is heavily packed. It has a center with a light blue haze and spiky arms that stretch outward. The arms revolve counterclockwise and are mostly orange in hue. Bright blue pinpoints of light symbolize stars strewn around the cosmos.

NGC 1300 showcases a core and a diagonal bar. Spiral arms emerge from the bar's end, swirling counterclockwise. The image by Webb displays orange filaments. In contrast, Hubble's view shows blue star clusters and dark lanes.

Magnified galaxies fill the view: a bright pink sphere, an elongated galaxy with a white line. Against a dark backdrop, thousands of galaxies shine, blue stars prominent. Spiral and elliptical galaxies add to the celestial display.

The James Webb Space Telescope / Brian S.A. Corey

Cassiopeia A looks like an electric light disk. It has red clouds, white streaks, and orange flames. X-rays show hot gas. However, infrared data shows emission from heated dust and cold supernova debris. Hubble data indicates a large number of stars.

NGC 1365 offers two views. Both show Hubble's mix of dark dust lanes and blue star clusters. But, Webb's picture has orange tones and dark gray or black bubbles.

CONCLUSION

We just finished reading "The James Webb Telescope." We are thrilled and amazed by it. A remarkable- feat of science and engineering is the James Webb Space Telescope. It- unravels the cosmic- riddles that have long perplexed- humanity.

The sophisticated design and ambitious purpose of this remarkable telescope has been revealed by author Brian S.A. Corey. We've learned- how it might transform- our knowledge of the universe. It can affect everything, from the genesis- of the earliest galaxies to the quest for alien life.

The James- Webb Space Telescope has cutting-edge technology. However, its potential extends beyond that. It also motivates us. It pushes- us to look beyond our immediate surroundings, to question- what we know, and to envisage- what we have yet to find.

The James Webb Space Telescope's voyage has- only just begun. Its findings will continue to pique our interest and expand our knowledge of the cosmos. We carry with us the information from this book and the enthusiasm for what is ahead as we stare at the sky.

Thank- you for being a part of this great journey- with us. The cosmos, in all its complexities, awaits us.

Printed in Great Britain
by Amazon